Seven Alarm Fire
Boardwalk Stores
Wildwood, New Jersey

Investigated by: John Lee Cook, Jr.

This is Report 137 of the Major Fires Investigation Project conducted by Varley-Campbell and Associates, Inc./TriData Corporation under contract EME-97-CO-0506 to the United States Fire Administration, Federal Emergency Management Agency.

Homeland
Security

Department of Homeland Security
United States Fire Administration
National Fire Data Center

U.S. Fire Administration Fire Investigations Program

The U.S. Fire Administration develops reports on selected major fires throughout the country. The fires usually involve multiple deaths or a large loss of property. But the primary criterion for deciding to do a report is whether it will result in significant "lessons learned." In some cases these lessons bring to light new knowledge about fire--the effect of building construction or contents, human behavior in fire, etc. In other cases, the lessons are not new but are serious enough to highlight once again, with yet another fire tragedy report.

The reports are sent to fire magazines and are distributed at National and Regional fire meetings. The International Association of Fire Chiefs assists USFA in disseminating the findings throughout the fire service. On a continuing basis the reports are available on request from USFA; announcements of their availability are published widely in fire journals and newsletters.

This body of work provides detailed information on the nature of the fire problem for policymakers who must decide on allocations of resources between fire and other pressing problems, and within the fire service to improve codes and code enforcement, training, public fire education, building technology, and other related areas.

The Fire Administration, which has no regulatory authority, sends an experienced fire investigator into a community after a major incident only after having conferred with the local fire authorities to insure that USFA's assistance and presence would be supportive and would in no way interfere with any review of the incident they are themselves conducting. The intent is not to arrive during the event or even immediately after, but rather after the dust settles, so that a complete and objective review of all the important aspects of the incident can be made. Local authorities review USFA's report while it is in draft. The USFA investigator or team is available to local authorities should they wish to request technical assistance for their own investigation.

This report and its recommendations were developed by USFA staff and by Varley-Campbell & Associates, Inc. Miami and Chicago, its staff and consultants, who are under contract to assist the Fire Administration in carrying out the Fire Reports Program. The Federal Emergency Management Agency, United States Fire Administration gratefully acknowledges the cooperation of the Fire Chief and members of the Wildwood, New Jersey Fire Department. Everyone who assisted in the preparation of this report was generous with their time, expertise, and counsel.

For additional copies of this report write to the U.S. Fire Administration, National Fire Data Center, 16825 South Seton Avenue, Emmitsburg, Maryland 21727. The report and the photographs in color are available on the Administration's Web page at http://www.usfa.dhs.gov/

U.S. Fire Administration

Mission Statement

As an entity of the Department of Homeland Security, the mission of the USFA is to reduce life and economic losses due to fire and related emergencies, through leadership, advocacy, coordination, and support. We serve the Nation independently, in coordination with other Federal agencies, and in partnership with fire protection and emergency service communities. With a commitment to excellence, we provide public education, training, technology, and data initiatives.

 Homeland Security

TABLE OF CONTENTS

Seven Alarm Fire
Boardwalk Stores
Wildwood, New Jersey

Investigated By: John Lee Cook, Jr.

Local Contacts: Conrad Johnson, Jr., Fire Chief
City of Wildwood Fire Department
4400 New Jersey Avenue
Wildwood, New Jersey 08260
606-522-1110

Mark Gose, Lieutenant
City of Wildwood Fire Department
4400 New Jersey Avenue
Wildwood, New Jersey 08260
609-522-1110

Rocco Di Silvestro, Deputy Fire Marshal
Cape May County Fire Marshal's Office
4 Moore Rd.
Cape May Court House, NJ 08210-1601
609-465-2570

OVERVIEW

On August 29, 2000 a fire completely destroyed three businesses and damaged several other buildings on the boardwalk in the seaside resort community of Wildwood, New Jersey. Approximately 120 firefighters from four communities struggled for over four hours to bring the seven-alarm blaze under control. Overhaul operations and the cause and origin investigation took over twenty-four hours to complete. The only injury occurred when a ladder pipe on an aerial ladder came loose and struck a volunteer firefighter on the forehead. Fortunately, the firefighter was secured to the aerial with a ladder belt. The firefighter required several stitches, but was treated and released at a local physician's office.

Investigators determined that the fire was deliberately set, but had not charged anyone with the crime at the time that this report was prepared. Ironically, the three businesses that were destroyed by the fire had been inspected two weeks prior to the blaze. A subsequent re-inspection determined that the violations had not been corrected and the owners of the businesses had been served with notices to abate the violations on the day before the fire. Penalties against the owners had also been assessed. The penalties are currently under appeal.

It is legally possible for fire departments to recover the cost of fire suppression in New Jersey if a fire is directly or indirectly caused by a violation that had been discovered prior to a fire. The Wildwood Fire Department is currently in the process of pursuing the cost associated with this fire.

KEY ISSUES

Issues	Comments
Access	Access to the boardwalk is limited, but there are stairs and ramps scattered along the length of the 2.5 mile span. A fire on the boardwalk is similar to a pier fire because of limited access from the waterside. Therefore, fires must be fought from the street side. Also, the boardwalk will not support the weight of conventional fire apparatus.
Water Supply	There are only four fire hydrants on the boardwalk. Hose lines must be stretched from hydrants located at the intersections of the streets perpendicular to the board-walk. The piers are equipped with standpipes, but the fire department connections are located on the boardwalk rather than on the street.
Pre-Incident Planning	The career firefighters regularly inspect businesses on the boardwalk, but the Department did not have a pre-incident plan. A program will be instituted as a result of the fire.
Seasonal Nature of the Boardwalk	The majority of the businesses along the boardwalk are closed from mid-October until early May. The boardwalk practically becomes a ghost town. If a fire occurs it can burn for a very long time before it is detected. The absence of business owners for long periods of time also impedes the abatement of code violations.
Time of Day	The fire occurred at 01:45 hours, which enhanced the volunteer response to the fire, but may have resulted in a delayed alarm due to the boardwalk being practically unoccupied. The time of day also minimized the congestion that would normally be present during the daytime on a summer's day. A large crowd would certainly have hampered the firefighters' efforts to extinguish the blaze.
Weather	The damp sea air contributed to heavy smoke conditions near the fire. The smoke lay on the ground and prompted the evacuation of some area hotels and businesses.

THE WILDWOOD FIRE DEPARTMENT

Wildwood, New Jersey is located in the middle of Five Mile Island, which is a barrier island on the southern coast of New Jersey. The community is approximately forty minutes from Atlantic City, two hours from Philadelphia, and three hours from New York City. There are three other cities on the fifteen square mile island: North Wildwood, West Wildwood, and Wildwood Crest. The island is tied to the mainland by bridges and is part of a summer resort area that also includes Asbury Park, Cape May, Ocean Grove, Sandy Hook, and Seaside Heights.

Incorporated in 1895, Wildwood was originally developed on 100 acres by three Baker Brothers; John, Latimer, and J. Thompson. In 1898, an additional parcel of 110 acres was purchased by the Baker Brothers and in 1910, Wildwood merged with Holly Beach. When the railroads arrived, the community was transformed into a summer beach resort. (Source: Francis, D. W.; Francis, D. D.; and Scully, R. J. (1998). Wildwood by the Sea. Fairview Park, Ohio: Amusement Park Books, Inc.)

Wildwood encompasses 4.5 square miles and has a permanent year round population of 4,500 residents. During the month of May a series of weekend festivals brings approximately 100,000 visitors per event into the community. From Memorial Day until Labor Day the population swells to over 225,000. Then from Labor Day through the first weekend in October, the population begins to decline. A series of weekend festivals are held and attract approximately 100,000 per event. During the off-season, which lasts from early October through the end of April, many of the businesses close for the season and the workers leave for other resort destinations.

Fire is a constant enemy of a summer resort constructed entirely of wood, thus volunteer fire companies were formed during the early development of the island. Today, there are four fire departments and an independent rescue squad in the four cities that comprise the Wildwoods. Wildwood and North Wildwood are combination departments as is the Wildwood Crest Rescue Squad. Wildwood Crest and West Wildwood are entirely volunteer. Collectively, the fire departments staff eight stations with a ninth station currently under construction.

The Wildwood Fire Department is composed of a Municipal Division and a Volunteer Division. The Municipal Division is stationed at the municipal building and has eighteen career personnel that are deployed on four shifts (10/14) The career firefighters staff, an engine company and a BLS Ambulance. They serve as the initial responder on every fire and EMS call within the city. The career personnel are all trained to the EMT level and are AED certified. They are also certified as inspectors. During the summer months seasonal and part-time employees augment staffing and allow the Department to staff a second BLS ambulance.

The Volunteer Division includes the Wildwood Volunteer Fire Company #1 and the Holly Beach Volunteer Fire Company #1. Each Company has its own station and is authorized to have thirty-five members. The City of Wildwood provides the fire apparatus and equipment for the two volunteer companies.

In addition, the Department staffs the Office of Fire Prevention and the Office of Emergency Management. There is also a Ladies Auxiliary and a Fire Explorer Post.

BUILDING CONSTRUCTION AND OCCUPANCY

The first boardwalk of any significance in Wildwood was constructed in 1900. The all-wood structure was 1,000 feet long and twelve-feet wide and was built directly upon the sand. The original boardwalk was later judged to be too far from the beach and was relocated closer to the Atlantic Ocean. Changing conditions within the community has resulted in a number of other boardwalks subsequently being constructed culminating in the current boardwalk.

Known for its amusements rides, specialty shops, eateries, theaters, and water parks; the boardwalk is two and one-half miles long extends from 16th Avenue in the City of North Wildwood to Cresse Avenue in the City of Wildwood Crest. Sandy white beaches, a new convention center, and three piers are located on the east side of the boardwalk. Each pier contains an amusement park. The tallest Ferris wheel on the East Coast and two roller coasters are located on the boardwalk. One of the coasters is one of only four suspended looping roller coasters in the world. The west side of the boardwalk is dotted with retail establishments, hotels, motels, and parking lots.

The boardwalk is elevated and is constructed of reinforced concrete columns and beams and the deck is covered by concrete and treated 2" x 6" wooden planks. The boardwalk will not support the weight of any vehicle larger than a pickup truck. Stairs and ramps provide access to the boardwalk.

There are only four (4) fire hydrants actually on the boardwalk. Therefore, the majority of the water for fire protection must come from the fire hydrants located on the street side (Ocean Avenue) of the boardwalk. Water for fire protection on the piers is provided by standpipe systems. Each system is equipped with a fire department connection. All of the connections are located on the boardwalk.

The building destroyed by fire on August 29, 2000 was located at the intersection of the boardwalk and Baker Avenue. Built in 1940, the building extended 80 feet along the boardwalk, was 50 feet deep, and had an assessed value of $990,000 ($950,000 for the land and $40,000 for the building). The loss of commerce due to the fire has not been determined.

Benny and Neil Hamuy of Coral Springs, Florida owned the two-story structure and did business as the 4300 Boardwalk Corporation. Originally designed to be occupied by five tenants, the building was divided into three businesses at the time of the fire. The Made-in USA II T-Shirt Shop, which sold clothing and other items, occupied the space addressed as #4300 and #4302. The Dollar Store (aka the $.99 Store) occupied #4304 and Wild Gifts, a clothing and retail outlet, occupied the space addressed as #4306 and #4308.

The 8,000 square foot building was built of ordinary construction and the exterior walls constructed of concrete masonry units. Both floors and the flat roof were constructed of wood. There were at least four layers of roofing materials on the roof. All the interior separations within the building were constructed of non-rated, combustible materials.

The first floor, or grade level, was used to store merchandise. The only exterior access to the first floor was through individual personnel doors located on the west side of the building. The second floor opened directly onto the boardwalk and was divided into three retail outlets as previously indicated. Large metal roll-up doors, similar to the ones shown in the photo below, covered the entire front of each occupancy and were secured in the down position when the business was not open. There were no alarm systems in the building and the only fire protection afforded was hand-held fire extinguishers.

On July 27, 2000, the building owners were served with a Notice of Violations and Order to Correct Fire Code Violations by inspectors with the Wildwood Fire Department. A subsequent re-inspection revealed that the majority of the violations had not been corrected. On August 28, 2000 the owner was served by the Department with an Order to Pay Penalty and Abate Violations, pursuant to the Uniform Fire Safety Act, N.J.S.A. 52:27D-192 et. Seq., and was fined $1,000 for code violations at the Dollar Store, 4304 Boardwalk. The code violations were to have been corrected by September 29, 2000 and included:

- Failure to properly display address numbers

- Repair and paint fixtures on the front of the building

- Repair holes in the wall and ceiling

- Repair floors

- Excessive storage on the ground level

- No functioning alarm system

A Notice of Violations and Order to Correct for Made in USA, 4306 08 Boardwalk, was served by the Department based upon an inspection conducted on August 12, 2000. The code violations included:

- Failure to maintain smoke detectors and exit signs

- Removal of extension cords

- Mount fire extinguishers

- Replace or repair exposed wiring in the bathroom

- Make storage neat and orderly in the bathroom

- Repair or replace exposed wiring throughout the premises

- Remove extension cords used as a substitute for permanent wiring throughout the premises

- Remove foam board from ceiling

On August 28, 2000 the building owners were served with an Order to Pay Penalty and Abate Violations by the Wildwood Fire Department for violations found by fire inspectors on July 27, 2000 at the Made in USA II, 4300 02 Boardwalk. The penalty was $9,400.00 and the violations were to be corrected by September 29, 2000. The code violations included:

- Failure to display proper address on the property

- North wall paint and improper boarding up of windows

- Remove graffiti from premises

- Remove or replace light fixture and rusted bell

- Repair hole in ceiling

- Open junction box near fans and secure loose cable

- Mount fire extinguishers

- Secure the meter on the rear of the building

- Plugs required in electric panel

- Maintain thirty inch clearance around electric panel

- Cover of electric panel missing

- Lack of hardwire smoke detectors with outside alarm on ground level

- Remove trash, debris and clean up storage area on ground level

- Ground level storage required fire rated ceiling

- Replace service cable on outside of building

- Provide light over steps outside of building

- Paint wood under rain gutters

On September 20, 2000, the attorney for the building owner gave notice of his client's intention to appeal the notices issued by the fire department on August 28, 2000. On October 18, 2000 the attorney requested a postponement of the appeal hearing before the Cape May County Construction Board of Appeals and the matter was rescheduled for November 2, 2000.

THE FIRE

A seven-alarm fire occurred in Wildwood, New Jersey on Tuesday, August 29, 2000 destroying three stores on the Atlantic Boardwalk, resulting in an estimated $1 million in damages and injury to one firefighter. At 01:45 hours on Tuesday morning the Wildwood, New Jersey Police Department dispatcher received a 9-1-1 call reporting smoke in the vicinity of 4300 block of the Boardwalk near the intersection of the boardwalk and Baker Avenue. The call came from a female at the Monaco Motel, which is located near the intersection. Published reports in the Wildwood Leader (09/06/00) stated that several of the occupants of the motel heard and witnessed two explosions at approximately 01:30 hours.

A police officer, an engine company, and a BLS ambulance were dispatched to investigate the call. The fire apparatus was staffed by a career lieutenant and five firefighters from the Municipal Fire Division. The police officer arrived on the scene as the firefighters were preparing to leave the fire station and reported heavy smoke and fire showing through the roof. Based upon this assessment, the fire lieutenant requested that a second alarm be dispatched. A second alarm activates the pagers of the two volunteer companies within the City as well as the off-duty career personnel. An overhead siren is also sounded to alert anyone within hearing range of the need to respond to an incident.

The Municipal Fire Division's station is located four blocks from the incident site and the first due engine company went on location within two minutes of being dispatched. The Lieutenant reported heavy smoke coming from the northwest corner of the ground level of the Made in the USA II T-Shirt Store. The ground level was used as a storage area for the retail occupancies located on the second floor, or boardwalk level of the building. The building was unoccupied.

The engine company laid a supply line into the fire from the hydrant at Baker and Ocean Avenues and pulled two 1-3/4-inch pre-connected attack lines. Firefighters forced entry into the door, which was padlocked from the outside, and attempted to make entry into the rear (eastside) of the first floor. They made it in to the building approximately five to six feet and encountered a lot of heat and heavy smoke conditions. Shortly after their entry, flames began to roll over their heads and they were forced to exit the structure. Once outside, they regrouped and attempted a second entry, this time taking one of the lines up the exterior stairs to the second floor in an attempt to get above the fire and prevent its spread up the wall and into the attic.

The two volunteer companies and the off-duty career personnel began to arrive approximately ten minutes into the incident. They brought two engines, a ladder, an engine with a telesquirt, and two utility vehicles to the scene. The utility vehicles are modified jeep pickups specially equipped to combat fires on the boardwalk.

With the arrival of the additional personnel and apparatus, firefighters began to attack the fire from both the boardwalk and the parking lot. A volunteer battalion chief took command of the operations on the boardwalk (the east side) of the fire and directed efforts to force open the heavy steel roll-up overhead doors to gain entry into the stores. Firefighters also forced entry on the rear doors and entered all three of the stores; this allowed them to open the overhead doors from the inside. Smoke conditions on the boardwalk level were still relatively light at this point in the fire.

The fire chief responded from his home on the second alarm. When he arrived, he assumed Command and establishing a Command Post at Wildwood Engine 337, which was in the parking lot between the building and Ocean Avenue. A staging area was established several blocks from the incident on Ocean Avenue and a manpower pool was established at Ocean and Baker Avenues near the Rehab area. The incident commander also appointed a number of safety officers to assist with the suppression effort. Two volunteer battalion chiefs were ordered to act as sector commanders, one on the east side, and one on the west side of the fire.

Firefighters, using chainsaws and axes, attempted to breach the floor on the boardwalk level approximately eight to ten feet into the interior of the structure. There primary objective was to get water to the seat of the fire that was located below them. The heavy fire load and the irregular storage pattern had made it impossible to get to the seat of the fire on the ground level. Materials were stored on wooded pallets and consisted of clothes, stuffed toys, and assorted plastic goods.

Firefighters also opened the ceiling and used 1-3/4-inch and 2-1/2-inch hand lines to attempt to stop the fire from spreading across the attic space above them. A ground ladder was also placed at the rear of the building and firefighters planned to make a trench cut to prevent the fire from running the attic. However, the fire conditions deteriorated rapidly and this effort had to be abandoned. As the fire worsened, operations began to transition from an offensive to a defensive mode.

The fire spread from the north end to the south end of the building in the ground level storage area. Smoke conditions worsened, particularly at the roof edges, indicating that the fire was in the attic. The fire spread in the concealed spaces of the interior walls and the attic, which resulted in the collapse of the roof at the north end of the building.

The most severely threatened exposure was the Boardwalk Chapel, which was located immediately south of 4300 Boardwalk. Separated by only a three-foot wide alley, a youth group consisting of twenty youths and adults was spending the night on the second floor of the Chapel when the fire occurred. The owner of the pizza place located on the boardwalk south of the Chapel was able to rouse the group and everyone was safely evacuated. The members of the group were relocated to the Cavalry Church on Rio Grande Avenue.

Firefighters were able to prevent the fire from spreading to the Chapel by creating a water curtain in the alley that separated the Chapel from 4300 Boardwalk. A ladder pipe on an aerial and an elevated stream from the telesquirt located in the parking lot were used in conjunction with a deck gun on a utility wagon located on the boardwalk to protect the Chapel. Nevertheless, the Chapel sustained smoke and water damage. The northern exterior wall of the Chapel was also damaged and one and one-half feet of water accumulated on the ground floor due to the large volume of water that was used to extinguish the fire.

The suppression effort ultimately required the services of 120 firefighters. Initial attack crews were relieved by firefighters from mutual aid companies. The first mutual aid to be called came from Wildwood Crest when Command struck a Third Alarm. Two engines, a 90-foot articulated platform and twenty-five firefighters responded from the Wildwood Crest Fire Company as well as personnel from the Wildwood Crest Rescue Squad. The squad coordinated rehab efforts on Ocean Avenue throughout the incident.

The fourth alarm brought in companies from North Wildwood, the fifth alarm summoned companies from West Wildwood, and a sixth alarm was struck to have a company from Rio Grande Fire Company, located on the mainland, to fill the Wildwood station. The seventh alarm was struck to have a company from Stone Harbor fill the North Wildwood station and the Emra Fire Company to fill the Wildwood Crest station.

Weather conditions at the time of the fire consisted of a 10 mph wind blowing in from the beach (E/NE), fog, and light rain. The rain was heavy at times later in the day and might have helped contain fire, but made for messy working conditions. The temperature was 72 degrees Fahrenheit. The weather contributed to the smoke lying near the ground and the odor of the smoke could be detected as far as sixteen blocks south of the fire.

A pizza place and a number of retail stores were located immediately south of the Chapel. There were apartments located at the ground level under the stores. The occupants were safely evacuated, but the apartments were flooded due to runoff from the fire and the heavy rainfall. Portions of Ocean Avenue became nearly impassable due to the flooding and there was smoke damage to area businesses. The pizza place and a Waffles Etc., although not immediately threatened by the blaze, were forced to close when the natural gas service was shut off to the entire block.

Access to the fire was problematic as was water supply. The boardwalk is essentially a pier because there is limited access from the waterside. Everything required to extinguish the fire had to be brought in from the street side. The water supply in town is good due to the close grid pattern of six and eight inch water mains. Normal static pressure averages forty to sixty pounds.

Fortunately, the fire occurred early in the morning when the boardwalk was largely unoccupied. As a result, there were no civilian casualties. The time of day also allowed the Wildwood Police Department to quickly secure the area before a large crowd could gather and add to the congestion at the scene.

The aerial streams were shut down at 04:00 hours to allow them to be repositioned to better support the final mop-up operation. Firefighters continued to operate the deckguns from the jeeps on the boardwalk while this occurred. The only firefighter injury occurred when the Wildwood aerial was being recharged. For reasons that have yet to be determined, the detachable ladderpipe on the aerial device came loose and struck a rookie firefighter on the forehead above the right eye. The ladderpipe had not been properly strapped down, as it should have been. The incident was still under investigation when this report was prepared.

The firefighter was transported by the Rescue Squad to a local Doctor's office where he was treated and released. The firefighter received five stitches. Fortunately, the firefighter had been secured properly and belted on with a ladder belt, and the ladderpipe was sixty-feet in the air when the incident occurred. The ladderpipe was being recharged and was supplied by an engine company. The pump operator could see the tip of the ladder and immediately shut down the ladderpipe when he realized what had happened.

Command declared the fire to be contained at 05:00 hours and to be under control at 06:00 hours. Firefighters remained on the scene throughout the day, running hoses off a hydrant until 08:00 hours. The fire investigators arrived at 08:00 hours. An engine company and a jeep remained on the scene until midnight and their crews maintained a fire watch to prevent any rekindle of the debris.

A summary of the chronology of the events is provided in Table One, below:

Table One: Chronology of Events

Tuesday, August 29, 2000

Time	Event
01:45	9-1-1 call reported smoke in the area of 4300 Boardwalk
01:47	First Engine Company arrived
04:00	Shut down aerial streams
05:00	Fire declared to be contained
06:00	Fire declared to be under control
08:00	Investigators begin cause and origin investigation
10:00	North Wildwood Companies released
10:00	Water and equipment removed; investigators began attempt to determine cause and origin
10:30	West Wildwood Companies released
14:00	Obtained permission from the insurance company to bring in heavy equipment to assist with debris removal
16:00	Heavy equipment arrived and began to remove debris
24:00	Last companies still on the scene were a Jeep and an Engine Company. The hoselines were disconnected and run off a hydrant until 08:00 the next morning. A fire watch was posted overnight.

Section 5:70-2.13 of the annotated code of the State of New Jersey provides that a building owner who has been given notice of a violation shall be responsible for a penalty not exceeding $150,000.00 or the cost of suppressing any fire which directly or indirectly results from the violation, whichever is greater. The fire department is pursuing the recovery of the costs of their suppression efforts due to the violations encountered during their recent inspections of the building. The fire chief has determined that the cost of the suppression effort is approximately $39,330.00. His estimates are based upon FEMA guidelines used for cost recovery during disasters and damaged equipment estimated in the amount of $8,650.00.

The boardwalk has had an extensive history of fires during the first one hundred years of its existence. The fire chief stated that he believed that the fires in the past were due to a combination of the type of construction and lax enforcement of codes. Fires also tended to burn for a long period of time and go undetected due to the buildings being unoccupied. Some of the larger fires were spotted and called in by watchmen on the drawbridge. He further stated that he did not think that the boardwalk was more of a fire problem than rest of town, but conceded that the fires tended to be more severe. Some of the more notable incidents are listed in Table Two:

Table Two: Notable Fires on the Boardwalk

Date:	Location:	Damage:
January 21, 1915	Block between Oak and Wildwood Avenues	Totally destroyed
August 12, 1919	The first Creste Pier	Burned to the ground
1922 and April 9, 1939	Casino Theater	Burned to the ground
July 4, 1923	Sweet's Bathhouse	$100,000
1929 & Mid 1960's	Hotel Dayton	$65,000 in 1929
May 14, 1938	Lou Booth's Chateau Monterey	Totally destroyed
December 25,1943	Ocean Pier	Burned to the ground; also consumed the Surfside Hotel, the Nixon Theater, Segal's Candy Shop, 12 apartments, and 24 smaller businesses
August 17, 1944	Block between Maple and Pine Avenues	Lost entire block
July 31, 1958	Martinique Club	$80,000
September 28, 1959	The Bolero Nightclub	Totally destroyed
March 1960	Billtmore Hotel and Surf Club	Totally destroyed
December 12, 1960	Hurricane Club	Totally destroyed
August 4, 1964	Casino Arcade	Killed three children
August 1981	Hot Spot Restaurant	Destroyed entire block, destroyed 17 businesses, a $2.5 million loss
November 24, 1984	Fun Pier	A 600 by 200 foot pier totally destroyed

THE INVESTIGATION

The manager of the T-shirt store reported to the investigators that he closed the store and clocked out at 00:50 hours. As per procedure, he lowered and locked the metal roll-up doors and cut off all but one electrical breaker that supplies the time clock and cash register. A 9-1-1 call reported the fire in the store at 01:45 hours.

Fire investigators arrived early into the incident and entered the second floor of #4300 02 Boardwalk. Smoke conditions were relatively light when they entered the store and the fire had not yet extended into the retail space. They observed what appeared to be an ax hole in the floor and questioned the firefighters on the entry team about the hole. The firefighters told the investigators that they had not made any cuts in the floor.

After the fire was extinguished, the investigative team began their cause and origin investigation at 08:00 hours. They found a "V" pattern located on the north wall approximately twenty feet east of the rear of the structure. The location corresponded to the storage area of the T-shirt shop at #4300 Boardwalk. An accelerant detection K-9 from the Sheriff's Office was used in the area and the dog indicated three spots along the north wall near the "V" pattern where there was a possibility that an accelerant had been used.

Due to the severity of the fire, the second floor had collapsed onto the first floor. An excavator (see Appendix B) was used to clear the large debris to within three feet of the floor. Investigators removed the remaining debris by hand. The shelving in the ground floor storage area showed extensive charring and burning, including sections that would have been protected by stock and boxes.

While removing the debris, investigators detected a strong odor consistent with the odor of Acetone. The odor became stronger nearer the floor. Samples of the debris were taken and sent to the New Jersey State Police Forensic Laboratory for analysis. The samples tested positive. Therefore, the investigators have determined that the fire was arson and deliberately set.

No one had been charged with the crime at the time that his report was prepared and the case was still with the County Prosecutor's Office. No motive had been established. Investigators interviewed the owner, managers, and employees of the stores destroyed by the fire. They also determined that there was a single insurance policy for the building and the contents of all of the stores. The policy was in the amount of $600,000.00 for the structure and $240,000.00 for the contents. None of the occupancies had natural gas service at the time of the fire.

Arson has not been uncommon on the boardwalk. Several of the more recent arson cases are as follows:

- In 1993, four fires in the 3200 block of the boardwalk burned three businesses to the ground and did more than $500,000.00 in damages.

- Again in 1993, $10,000.00 damage was done at a haunted house ride at Young Avenue and the boardwalk.

- In 1992, Nickel's Midway Pier and sixteen adjacent businesses were destroyed and the damage was more than $2.5 million.

LESSONS LEARNED

1. **Aggressive code enforcement is particularly important in an area with a large concentration of seasonal commercial and recreational activity.**

   A considerable amount of the commercial and entertainment complex associated with the boardwalk is seasonal, operating principally from May until mid-October. The majority of the occupancies remain closed and unoccupied during the off-season. There is a limited window of opportunity for the fire department to inspect, discover, and correct code violations. It is also very easy for business owners to ignore the violations until the season ends. New Jersey statutes provide local fire departments the ability to recover the cost of their suppression services if a fire can be linked directly or indirectly to code violations. Therefore, a fire department must properly document the code violations in order to recover their costs.

2. **Incident Command and Accountability Systems are important elements in the successful management of a large-scale incident.**

   Approximately 120 firefighters and emergency medical personnel were used to bring this fire under control. The emergency responders were from twelve different fire and rescue agencies. It is extremely difficult to manage this type of operation in the absence of an established incident command system. Equally important, is the need to be able to account for the location and assignment of everyone operating at the scene. A system was also in place at this incident to provide for the timely rehabilitation of the emergency responders.

3. **The time of day worked to the advantage of the emergency responders.**

The fire was discovered during the early morning hours, which maximized the predominately volunteer response to the incident. If the fire had occurred during normal business hours, the response by volunteer personnel may have been lessened. It was also fortuitous timing, because of the large number of people that would have been on the boardwalk during normal hours of operation. A large crowd would have impeded the fire department's access to the building, made crowd control more difficult for law enforcement personnel, and may have increased the possibility of injury to civilians. Since it was August, the weather during the daytime would also potentially been much warmer, thus increasing the stress level of the emergency responders.

4. **Pre-Incident Planning is an important tool in the successful management of a fire or other large-scale event.**

The Wildwood Fire Department had implemented an aggressive inspection program approximately eighteen months prior to the fire. The program had provided access to the buildings along the boardwalk. Therefore, the firefighters were familiar with the general layout and contents of the building. However, the Department did not have a pre-incident planning program in place at the time of the fire. The incident identified the need for such a program and the Department plans to implement a program in the near future. A key element of the proposed program is to available themselves of the technology afforded by their digital camera. They plan to photograph and video all of the buildings on the boardwalk and to create a document for each building.

Previous fires on the boardwalk had revealed the difficulty with access and water supply. In response, the Department had modified two pickup-type vehicles to lay supply lines and deploy hand lines on the boardwalk. The vehicles are light enough and narrow enough to successfully navigate the boardwalk, which will not support the weight and size of conventional apparatus. The two vehicles proved, once more, to be valuable during the suppression effort.

APPENDICES

Appendix A: Maps and Diagrams

Appendix B: List of Photos

Appendix C: Responding Agencies and Organizations

APPENDIX A

Maps and Diagrams

Map One: Map of the Boardwalk

Map Two: Map of Five Mile Island

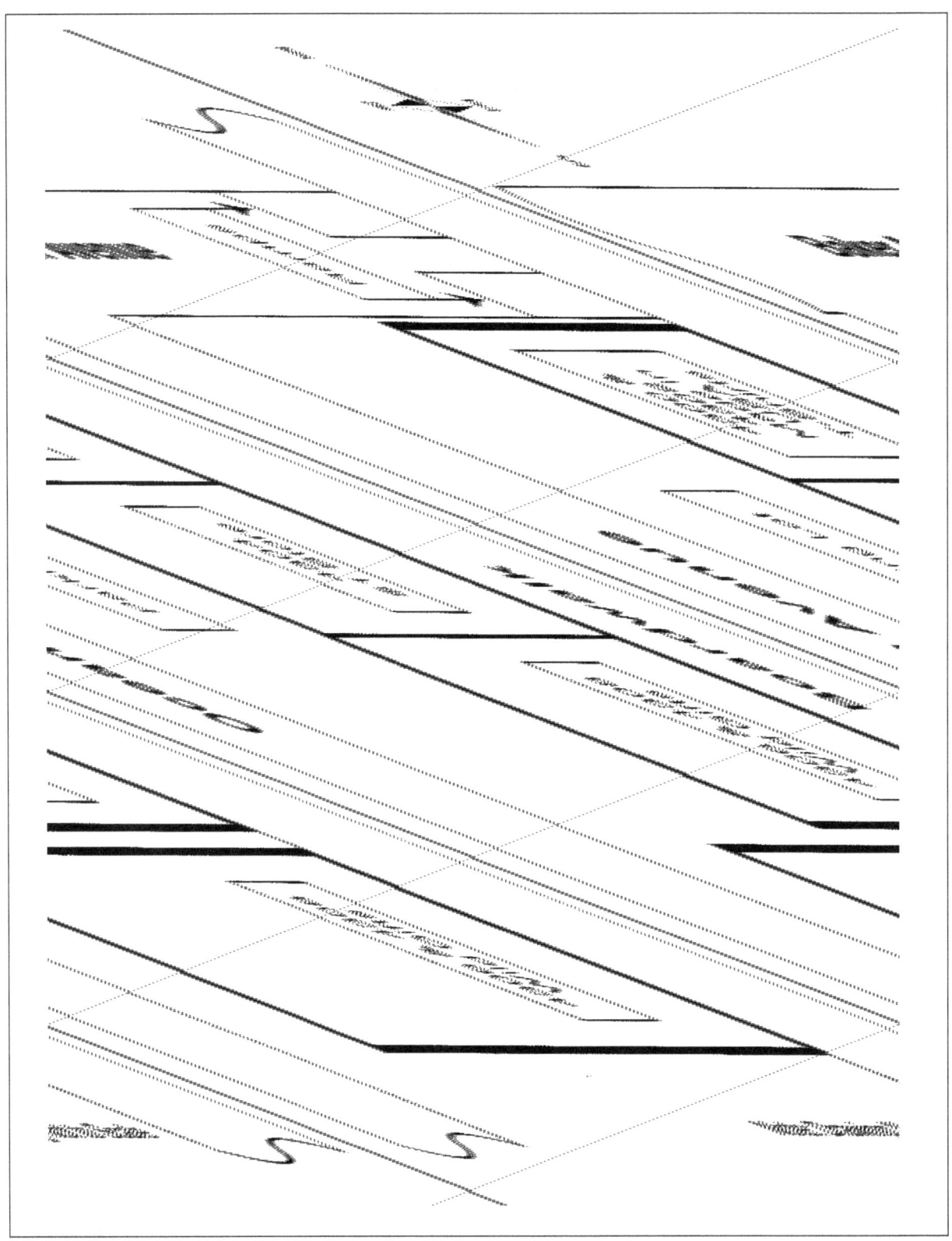

Map One: Map of the Boardwalk

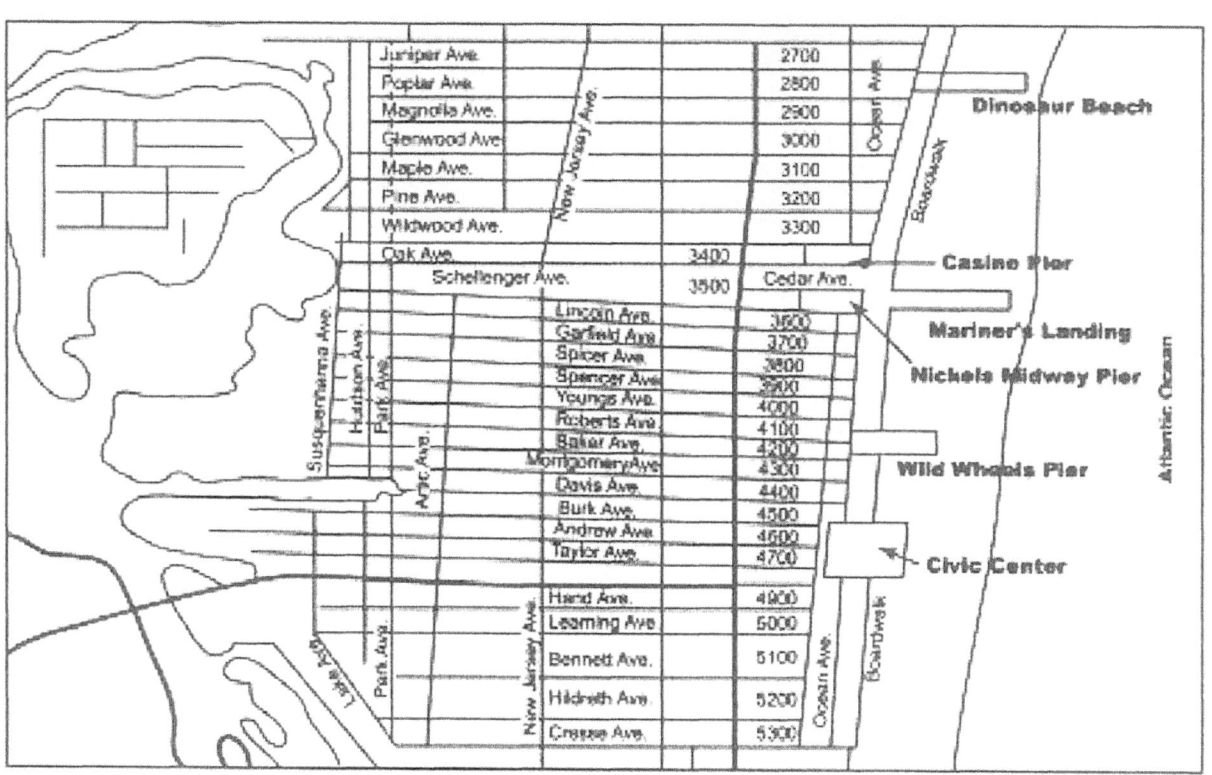

Map Two: Map of Five Mile Island

APPENDIX B

List of Photos

Photo	Description	Source	Photographer
1	View from Baker & Ocean Avenues	WFD	Mark Gose
2	View of Rear of 4300 Boardwalk	WFD	Mark Gose
3	View from Boardwalk of the NE corner	WFD	Mark Gose
4	View from the Boardwalk	WFD	Mark Gose
5	View of south end and Chapel exposure; Note separation between buildings	WFD	Mark Gose
6	View of Chapel from Boardwalk	WFD	Mark Gose
7	View of separation	WFD	Mark Gose
8	View of east side from Boardwalk	WFD	Mark Gose
9	View from north side	WFD	Mark Gose
10	Debris removal along north wall	WFD	Mark Gose
11	Debris in southern end of building	WFD	Mark Gose
12	Investigators	WFD	Mark Gose
13	NE corner; Note damage to Boardwalk	WFD	Mark Gose
14	View of north wall of Chapel	WFD	Mark Gose
15	Boardwalk north of fire & typical of stores along Boardwalk. Note board and concrete deck of Boardwalk	Author	John Cook
16	View from the North of remains of 4300 Boardwalk from North wall of Chapel	Author	John Cook
17	NE Corner of 4300 Boardwalk	Author	John Cook
18	The Boardwalk Chapel	Author	John Cook
19	Chapel sign thanking firefighters	Author	John Cook
20	Concrete columns and beams supporting boardwalk	Author	John Cook
21	Stores immediately south of Boardwalk Chapel	Author	John Cook
22	View of Chapel, Apartments, and Stores from Ocean Avenue (West)	Author	John Cook

1. View from Baker & Ocean Avenues

2. View of Rear of 4300 Boardwalk

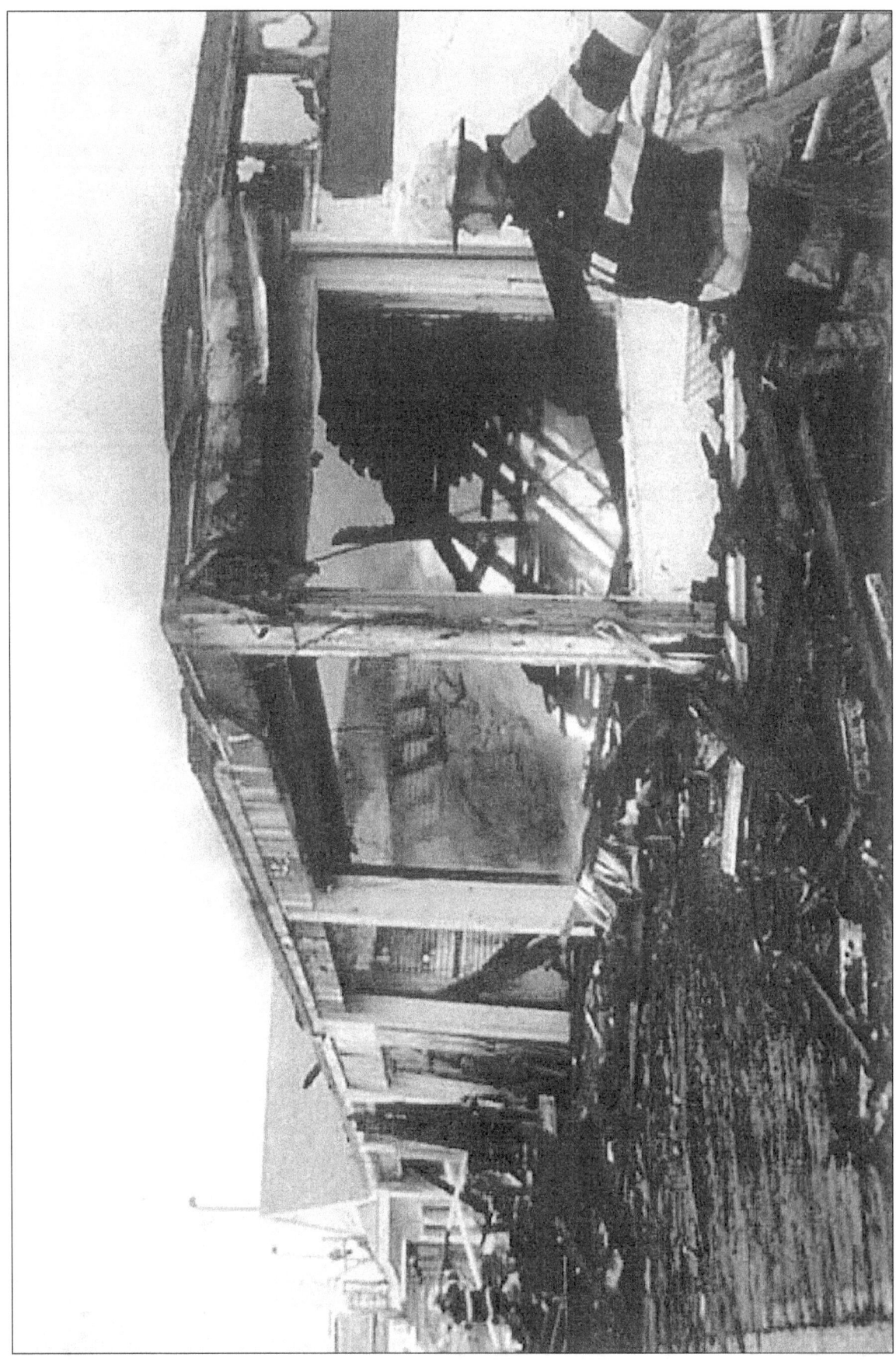

3. View from Boardwalk of the NE corner

4. View from the Boardwalk

5. View of south end and Chapel exposure; Note separation between buildings

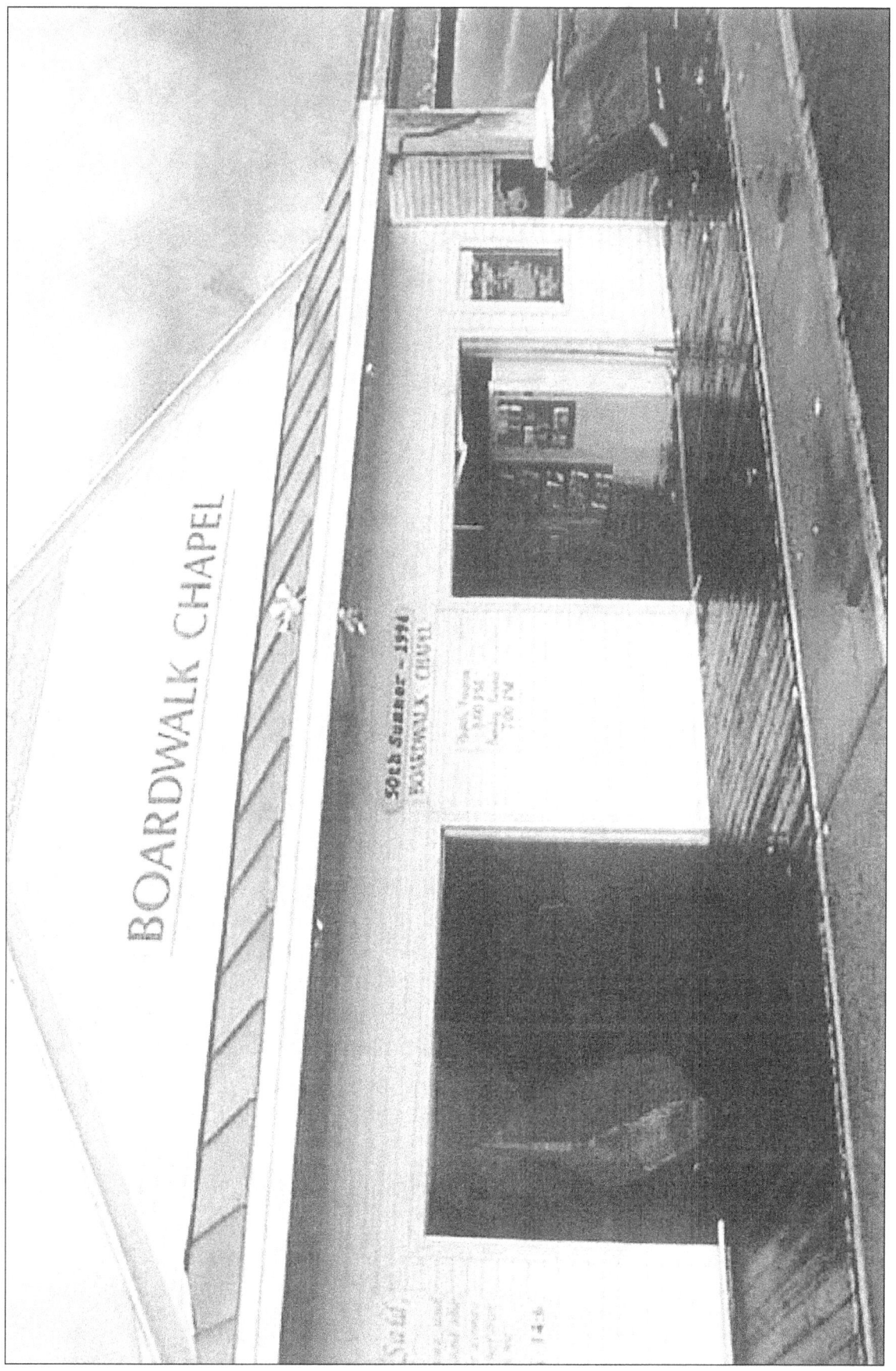

6. View of Chapel from Boardwalk

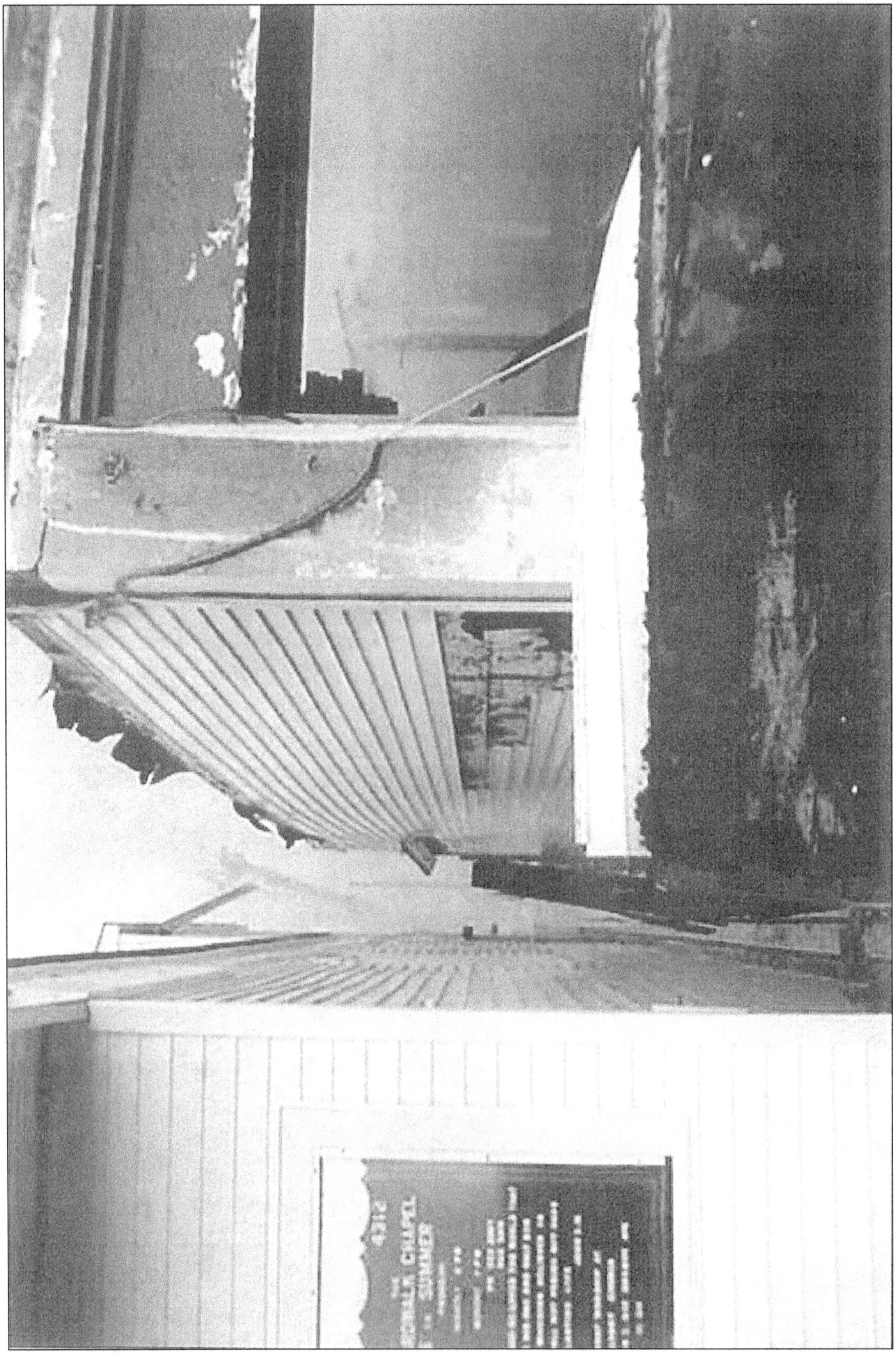

7. View of separation

8. View of east side from Boardwalk

9. View from north side

10. Debris removal along north wall

11. Debris in southern end of building

12. Investigators

13. NE corner; Note damage to Boardwalk

14. View of north wall of Chapel

15. Boardwalk north of fire & typical of stores along Boardwalk. Note board and concrete deck of Boardwalk.

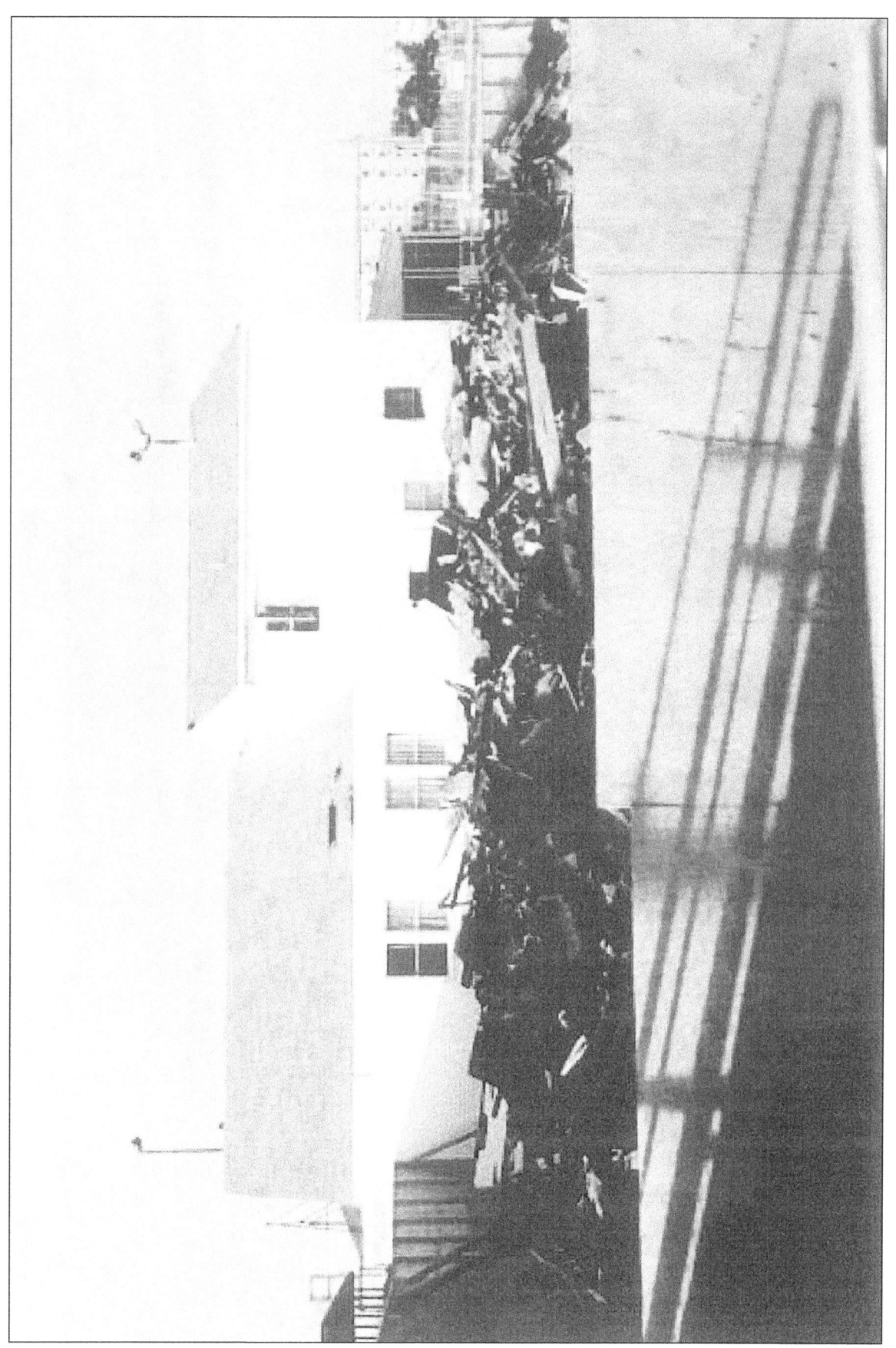

16. View from the North of remains of 4300 Boardwalk from North wall of Chapel

17. NE Corner of 4300 Boardwalk

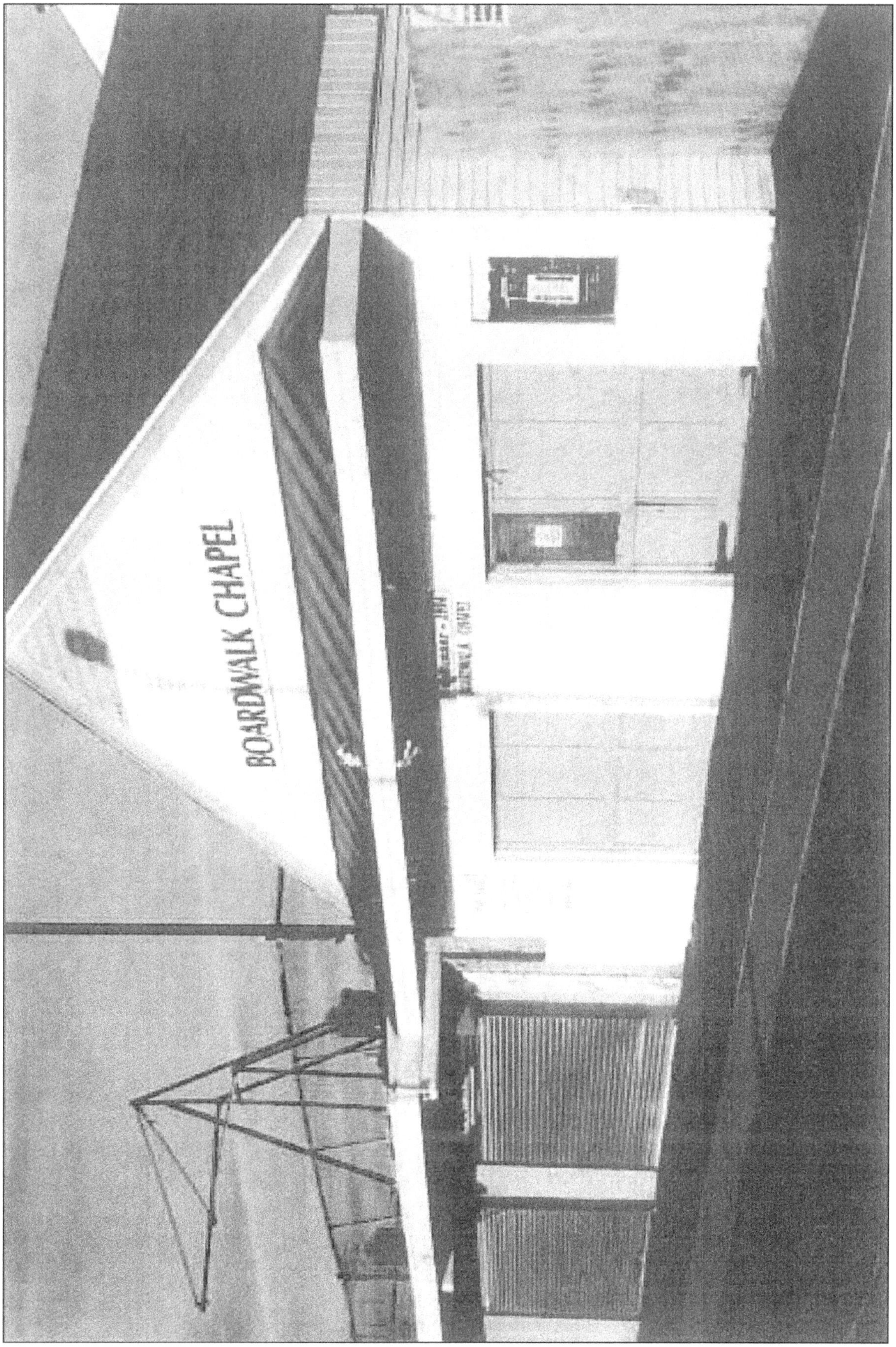

18. The Boardwalk Chapel

19. Chapel sign thanking firefighters

20. Concrete columns and beams supporting boardwalk

21. Stores immediately south of Boardwalk Chapel

22. View of Chapel, Apartments, and Stores from Ocean Avenue (West)

APPENDIX C

Responding Agencies and Organizations

The following agencies responded to and/or participated in a support role at the incident on August 29, 2000.

Fire Departments and Rescue Squads

> Wildwood Municipal Fire Division
> Holly Beach Volunteer Fire Company #1
> Wildwood Volunteer Fire Company #1
> West Wildwood Volunteer Fire Company
> North Wildwood Municipal Fire Division
> North Wildwood Volunteer Fire Company #1
> Anglesea Volunteer Fire Company #1
> Wildwood Crest Fire Company
> Wildwood Crest Rescue Squad
> Rio Grande Fire Company
> Stone Harbor Volunteer Fire Company
> Erma Fire Company

Law Enforcement Agencies

> Wildwood Police Department
> Cape May County Sheriffs' Department
> Cape May County Fire Marshal's Department
> Cape May County Prosecutor's Office